ALBERT EINSTEIN
A MAN OF MANY FACES

THE HISTORY HOUR

CONTENTS

PART I
Introduction ... 1

PART II
A GENIUS SHOWS UP WITH A DEFORMED HEAD
The Odd Shaped Head Starts to Read ... 7

PART III
Einstein Had A "Miracle Year" ... 25

PART IV
EVEN A GENIUS MAKES MISTAKES
Sometimes, the Mistakes Come Back to Haunt the Family ... 37

PART V
Einstein Finds He Has Enemies ... 45

PART VI
Albert Had His Problems Too ... 55

PART VII
Did Einstein Have a 3rdSon? You Decide ... 63

PART VIII
The End is Soon to Come ... 69

PART IX
Chapter 8: What Exactly Was the Legacy of Einstein? ... 77

PART X
CONCLUSION
Strengths ... 83
Weaknesses ... 84
Interesting Books on Albert Einstein you may enjoy: ... 86

Copyright © 2018 by Kolme Korkeudet Oy

All rights reserved.

No part of this book may be reproduced in any form or by any electronic or mechanical means, including information storage and retrieval systems, without written permission from the author, except for the use of brief quotations in a book review.

❦ I ❧
INTRODUCTION

※

Most readers do not realize that Albert Einstein was a man of many faces. He overcame many obstacles in his childhood to become the person he was as an adult.

※

That being said, I will let you be the judge as to the person he became as an adult. We all know he was a pure genius, but what kind of person was he outside of his world of genius?

※

Even though he was a man that not only wanted peace in the world, he still created **a *formula/theory*** *that would build* **a bomb** *capable of killing thousands at a time.*

His formulas have led us to such modern technology that we might not have today in the palm of our hands and changed the way we live our lives.

Through *Albert's breakthroughs in physics*, he has made it possible for us to have items such as televisions, global positioning satellites, computers, CD players and millions of more things too numerous to mention that are at the touch of our fingertips every day.

He has helped bring forth inventions that have simplified our lives more than anyone ever dreamed of at the time of his discoveries.

As far as the way he lived his life, well, read on, and you be the judge to see if you think he had a full and happy life. Einstein's story awaits you on the pages ahead.

II
A GENIUS SHOWS UP WITH A DEFORMED HEAD

"Few are those who see with their own eyes and feel with their own hearts."
ALBERT EINSTEIN

❦

The place was Württemberg, Germany and the date, **March 14, 1879**, at 11:30 AM when a baby boy was born to *a featherbed salesman and his wife*. No one at that time had any inkling that one day this baby would stand the world on its head. The child's birth weight cannot be found recorded anywhere in history and for sure not on his birth certificate. That more than likely could be attributed to the fact that few had a means of weighing a newborn at a home birth back in those days.

❦

This town today has a population of about 120,000 and a small plaque is all that is left to commemorate where Albert Einstein's home stood when he was born. The house was destroyed during World War II.

Albert's mother, Pauline, noticed after his birth that his *head seemed unusually large*. She could not help but worry about it and could not get it out of her mind. She noticed that it was oddly shaped, mostly on the back side.

One thing that upset both parents was the way he was developing intellectually. His speech was so delayed that he did not begin to speak until he was 3 or 4 years old and not fluency until he was nine. Even with lack of speech development, he still managed to be one of the top students at the elementary school he attended.

Looking back, researchers at Oxford and Cambridge Universities feel that Albert showed signs of *one form of Autism called Asperger's Syndrome*.

Many who have been diagnosed with Asperger's are thought to be eccentric. They seem to lack social skills and seem to be obsessed with topics that are complex, and they may have issues communicating with others.

❦

Researchers think that Einstein, from a young age, exhibited signs of Asperger's.

❦

There were signs when he was a child. For instance, the way he would pull away from others and be such a loner and how *he would repeat sentences over and over*. Albert would keep up the repeating until he turned seven years old. The Asperger's even reared its ugly head later when he became a professor and gave lectures. So much was going on in his mind at once that it seemed to confuse what would come out of his mouth.

❦

Researchers also say that his passion, standing up for what he thought was justice, falling in love; all are exactly compatible with Asperger's. But, what most individuals that suffer from Asperger's can't do is small talk. So, if he found the right woman that could mingle well in social situations, he would not have to chit chat at social events as she could do that for him.

❦

When Einstein grew older he made intimate friends, he spoke out on about every political issue you could imagine, and he had numerous affairs even though he was married.

❦

His mother's maiden name was Pauline Koch, and she made

sure the families household ran like a well-oiled machine. Albert was joined two years later by a sister who was named Maria.

❦

When Albert was five years old, his uncle gave him **a compass**, and it immersed him in the fact that *an invisible force could make that needle move*! That compass alone at age five was what would start him on his journey through life with anything that had to do with invisible forces.

THE ODD SHAPED HEAD STARTS TO READ

❦

His father later achieved with moderate success the oversight of *an electrochemical factory* after the family moved to Munich.

❦

While living in Munich, Albert turned twelve years old; he happened upon a book called **"geometry,"** which he lived with his nose inside of day and night. He often referred to it as his "sacred geometry book."

❦

At twelve, he also became very religious. He wrote many songs of praise about God and would chant gospel songs on his way to school. It started changing after he began reading some of the science books that were in direct contradiction to his beliefs about religion thus far.

While attending school at the Luitpold Gymnasium, Albert felt out of place and bullied, as we call it today, most of the time because of the education system used by the Prussians that meant to hold back those who seemed to be creative and original in their thinking. One teacher at the Gymnasium went so far as to tell Albert he would never turn out to be anything.

Albert *learned to play the violin*, and he was quite good at it, and he enjoyed playing it. He had never become interested in music until he had heard the music of Mozart and then he was hooked. He was a fan of Beethoven and Bach, but **Mozart** was by far his favorite. It was said that playing his violin or the piano helped him when he was working on his theories. *He could play a few notes on the piano or the violin, and then he would jot down notes on some theory.*

But, thank goodness for a young medical student by the name of **Max Talmud** who often came to their home to dine with his family. Talmud started out by working as a sort of informal tutor to Einstein on a higher level of learning in philosophy, science, and mathematics. Talmud made Albert's fascination with all things regarding physics even more heightened.

One very pivotal moment in Einstein's life was at age 16

when he read **"Popular Books on Physical Science,"** wherein the author himself talked about being able to imagine riding so fast that you could keep up with what was traveling inside a wire. Einstein could barely think of anything else for the next ten years. What would it be like if you could run beside a beam of light at the same speed? If perhaps you could do this, it would seem as the beam of light was standing still.

※

During this time, Albert had written his very first of many **"scientific papers,"** (The Investigation of the State of Aether (what we know as ether) in Magnetic Fields").

※

Albert's father never seemed to be able to get anywhere with any of his businesses. In 1894, his company lost a contract they needed to stay operational, so *the family had to move to Milan* for his father to work with one of his relatives.

※

They left Albert by himself at a boardinghouse in Munich and told him to finish his education there. Albert was miserable and despondent, so alone and utterly repelled by the idea that he would be required to serve military duty when he turned 16.

※

Einstein ran as fast as he could six months later and wound up on his Mother and Father's doorsteps. They were shocked to see him and anxiousabout what he would be facing as **a**

draft dodger, school dropout, and he had no employable skills. What in the world were they going to do with him?

※

But, desperate times do call for desperate thinking and drastic measures and out of that can come some great answers. It was with great luck that *Albert was able to apply to The Swiss Federal Institute of Technology*, even if he did not have a high school diploma, if he could pass the tough entrance exams.

※

Albert excelled in physics and math, but *he failed at chemistry, French, and biology*. Because Albert had such exceptional math scores, they allowed him to go ahead and enter the school on one condition, and that was if he first finished formal school.

※

Albert attended a "special" (a progressive and free-thinking) high school **in Aarau, Switzerland** which was run by **Jost Winteler** and graduated within three semesters in 1896. While attending there, he lived with Jost Winteler and his family.

※

Albert had begun what would become a lifelong friendship with the entire Winteler family while living with them during the three semesters. *Albert's first love was Winteler's daughter*, **Marie**. Albert's baby sister, Maja, would later marry the son of Winteler, Paul.

❦

At times when Einstein looked back on his life, he would have to say that his happiest years were those spent in Zurich. He met many loyal friends and lots of students. It was in Zurich that he met his first wife **Mileva Maric** who was a physics student from Serbia.

❦

Mileva was the only woman pursuing a physics major in Zurich while Einstein was attending the college. It was during the second semester that the two became interested in each other.

❦

Einstein's mother was against Mileva from the beginning. It did not keep Albert from romancing Maric while the whole time his Mother and Dad were not happy with what looked like "soon to be a wedding." His Mother did not like her because she had a Serbian heritage and her religion was Eastern Orthodox Christian.

❦

After Einstein graduated from Polytechnic, he took a job away from Zurich, while Mileva had to stay there in Zurich. Mileva was not doing well with her grades as she failed her final exams twice.

❦

Albert had convinced Mileva to meet him at *Lake Como* for a romantic weekend vacation. It was a well-known resort that had beautiful scenery of snow-capped mountains. It was this weekend when Mileva conceived their illegitimate child.

༺✦༻

Einstein would come to see Mileva every Sunday to visit. It was when Albert came during one of those visits that Mileva told him she was pregnant.

༺✦༻

Meliva quit school because she had failed her physics exams two times. She was so depressed that she went to Hungary to her parent's home where she had to tell them about both of her failures, facing them all by herself. In the beginning, her father said she could not marry Albert.

༺✦༻

Then came the winter of 1902, Mileva delivered the baby girl that they had named **Lieserl**. Mileva had a difficult time during the birth process, and Einstein was not there to support her. He found out about the baby through a letter from Mileva's father. Einstein was so excited about his new child that he wrote back full of questions. He wanted to know if she cried adequately, if she was healthy, who she looks like. He was so happy and so in love with the baby and had not even seen her yet.

༺✦༻

In the letters between Mileva and Einstein before Lieserl's birth, Einstein had told her to nurse the child instead of giving the baby cow's milk. He was afraid the cow's milk would make the baby stupid.

❦

It was in 1902 that Einstein was lower than the lowest in his life. He was not able to marry Maric because he didn't have a job, and therefore, could not support a family. His father's business went belly up again. Einstein was unemployed and desperate and finally decided to take one of the lowliest jobs you could get, and that was tutoring children. As luck would have it, he even got fired from tutoring.

❦

It was about that time that Albert's father took to *his deathbed*, and right before he passed away, he told Albert he had *his blessing to marry Maric/Mileva*.

In the late part of 1902, a bit of good fortune came Albert's way. A friend of his father, Marcel Grossman got him on at the Swiss Patent Office in Bern as a clerk of sorts.

❦

For some reason, despite all his interests, Albert never traveled to meet his new infant. No one knows why. It is where the story about the baby starts to get darker. Maric moved back to Zurich to wait for Albert to get the job at the patent office so they could get married, but she did not bring Lieserl with her. *Albert and Maric married in January 1903*, and they moved into an apartment.

It did not keep Albert from romancing Maric while the whole time his Mother and Dad were not happy with what looked like that might soon be a wedding. His mother did not like her because she had a Serbian heritage and her religion was Eastern Orthodox Christian.

More curious than ever is in all the research that has been conducted and all the letters back and forth between Albert and his wife there seems to be no common thread running through the stories of the time. August 1903, Maric found out that *Lieserl who was about 18 months was sick with scarlet fever*. She went to see the baby. In September, Einstein wrote to Maric who was still there with the baby. He was worried about their little girl. He asked her if Lieserl was registered. He told Maric they had to be careful so that it did not cause the child problems in her future.

No one knows if the tiny child died of Scarlett Fever or maybe they gave her up for adoption.

Giving her up for adoption seems far-fetched when you realize the streak of stubbornness Einstein wore all the time. He never gave in or gave up on what lay before him. In that period, it was common that children died of scarlet fever and as young as the baby would have been, it could very well be a definite possibility. It shall forever remain a secret to all unless somewhere a letter that has never seen the light of day comes forth that reveals **this secret**.

※

While Maric was away, she realized she was pregnant again, and this time it was a boy, **Hans Albert Einstein**.

※

Nothing can be found as to what happened to Lieserl. So many reporters, journalists, and researchers have searched for this missing child of such a brilliant man, and yet, she has disappeared without a trace.

※

Albert did not seem to be a perfect husband, but it must be said that he did try in the beginning. He was just so into his work and with all of his attention on his work that it was like he did not even know his family was there. It was like he was not "in the present" with the rest of his family.

※

Albert dug into his work even more. Mileva sank into depression. (One must wonder if Mileva did not suffer from postpartum depression as it had not yet been discovered at that time.) Their house, seen by one visitor, was an absolute mess. It is said that Albert did try to help, but he couldn't get into it. He would pick the baby up and push the stroller around and be writing equations on his notepad all at the same time; he hardly realized he was a father and had father duties.

❦

July 28, 1910, **Eduard Einstein** was born. Things seemed to improve for them for a while, but it didn't last. Mileva was still depressed, or maybe she just had post-partum depression again, who can tell by the writings that have been found and she was getting jealous of all those women that Einstein openly flirted with in front of her or those he bragged about to her face.

❦

In 1910, Albert's mom moved to live with her sister Fanny and the rest of Fanny's family to Berlin after Albert's father had passed away. Now bear in mind, **Elsa** (*Albert's second wife*) was Fanny's daughter. Pauline took a job as a housekeeper in 1911.

❦

She eventually moved to live with her brother, Jacob Koch, in Zurich in 1914. During the days of World War I, Pauline become ill with cancer.

❦

Einstein and his little family moved to Prague in 1911, where he was going to teach at the university. Meliva did not like living in the city.

❦

One year later, Albert had an offer from Zurich, and they moved there. Mileva was so happy to be in Zurich. They were there only a couple of years though, because *in 1914, Einstein was offered a position at the University of Berlin* and again they moved.

❦

Mileva hated moving there and was very unhappy. Albert's cousin, Elsa, would be close at hand for him. Mileva was very jealous of her.

❦

And Mileva's suspicions were spot on; Albert started dating Elsa. It started the end of Mileva's and Albert's marriage.

❦

The marriage kept deteriorating, and they attempted to glue it back together for the children's sake. Einstein, as a pacifist, sat down and wrote a "list of conditions" that his wife had to accept when he got home if they were going to stay married. The following is the verbatim list he gave her.

❦

CONDITIONS:

- *You will make sure:*
- *That my clothes and laundry are kept in good order;*
- *That I will receive my three meals regularly in my room;*
- *That my bedroom and study are kept neat, and especially that my desk is left for my use only.*
- *You will renounce all personal relations with me insofar as they are not completely necessary for social reasons. Specifically, You will forego:*

- *My sitting at home with you;*
- *My going out or traveling with you.*
- *You will obey the following points in your relations with me:*
- *You will not expect any intimacy from me, nor will you reproach me in any way;*
- *You will stop talking to me if I request it;*
- *You will leave my bedroom or study immediately without protest if I request it.*
- *You will undertake not to belittle me in front of our children, either through words or behavior.*

※

This list is chilling and to think that he would make such demands of the womanhe had been so in love with in his early years makes it hard to comprehend. What had happened to them as a couple?

※

I do realize that it is not fair to judge someone this far back in time and not know all the details of both sides, but it is hard to read this and not feel chills down your spine.

※

Einstein and his wife were continually fighting now about the children and howeager their finances seemed to be all the time. Albert was sure his marriage was over and began an affair with his cousin, Elsa.

※

At first, his wife accepted the list he gave her, but within three months she said enough was enough and Mileva and the boys returned to Zurich. Einstein was reported to have been standing on the platform waving goodbye and crying, the reason he was crying, no one knows. Because in just a few weeks he seemed to be over it and was as happy as a lark living alone in what he called "his tranquility."

※

In 1918, while Pauline was visiting Maja, her daughter and Paul Winteler, Maja's husband in Luzern, they realized how ill Pauline had become and took her to Rosenau, a sanatorium where she could live and receive care.

※

When Albert asked Mileva for the divorce, she had a nervous breakdown, and it seemed she was slow to recover.

※

He had finally divorced Mileva in 1919 and agreed with her that if he ever won the Nobel Prize, he would give the money he might receive from it to her as part of the divorce settlement.

※

At the end of 1919, Albert came and removed *his terminally ill mother* from the sanatorium and moved her in with Elsa and himself in Haberlandstrasse where she passed away a few months later in 1920. But, at least she was with family when she died.

❦

Then their son Eduard became a worry for him and Mileva. Eduard was very gifted. He was reading Schiller and Goethe in first grade and was blessed with a photographic memory. He could learn whatever he set his mind to at breathtaking speed. With all his intelligence Eduard was troubled. (Eduard had to finally be admitted into a mental facility due to schizophrenia, and *he died there in 1965.*)

❦

As Einstein had promised Mileva when **he won the Nobel Prize for Physics in 1922**, he would give the money from it to her. But he kept the award and gave proceeds equaling almost ten times what an average professor made to Maric. Maric was smart and invested it in three Zurich apartments, which she kept up until nearly the time of her death. (I have given you these two case scenarios of the money in the divorce settlement. Both came from reliable sources, and both seem plausible. However, from Albert's irrational acts at times, one tends to lean toward Albert losing most of it in the stock market crash.)

❦

Mileva devoted a lot of her time in taking care of Eduard and in 1947, she started to deteriorate healthwise. In 1948, she suffered from a stroke that resulted in paralysis on one side of her body. *Mileva died August 4, 1948.* Eduard, as said before died in 1965. One wonders if he was ever visited by his brother Hans Albert Einstein or his father, Albert Einstein.

❦

Elsa being Einstein's cousin caused everyone to talk, but Einstein did not care. Elsa's dad was Rudolf Einstein, and he was the "rich uncle."

※

Elsa had previously been married to a Max Loewenthal, from Berlin and they had two daughters, Margot and Ilse. They also had a son who had died shortly after birth.

※

Albert and Elsa moved in together in September 1917. Elsa was the one who kept putting the pressure on Albert to finalize the divorce with Maric.

※

Albert's main attraction with Elsa was that *she was a great cook*. He was also grateful to her for taking care of him when he was so ill with stomach issues. There was never any passion between the two of them, or that is what most thought. They married on June 2, 1919. Elsa was 43 and Einstein was 40.

※

Albert's life makes you wonder if he was a narcissist because his personal life seemed to always be in chaos and he was so aloof and callous. Elsa's daughter **Ilse** sent Dr. George Nicolai a close friend of hers a letter and told him when he read the letter to tear it up, but apparently, he did not, because it still exists.

※

Remember, Albert had been in the throes of a divorce from Mileva so he could marry Elsa, his cousin. Ilse was the oldest of Elsa's daughters, and she was working as Einstein's secretary.

※

The letter was from Ilse, and it was a plea for advice. She told how she and Albert had been joking one day and all of sudden it got serious and Einstein asked her to marry him instead of her mother. She said that Albert told her he loved her, Ilse.

※

And, crazy as it was, her mom, was willing to move over and let Ilse marry Einstein if that would make her happy.

※

Albert was not going to make a choice, he said he would marry either one, he did not care. Ilse said that she knew he loved her very much, and probably more than any man, she would ever find because he told me so.

※

Ilse did not feel like that about Albert. She thought of him as a father figure and did not want any physical relationship with him whatsoever.

※

There's no evidence that Ilse and Albert ever consummated

their relationship. The next year Albert and Elsa were married and stayed married until *Elsa died in 1936*.

❧

Ilse married Rudolf Kayser who was a literary critic and a writer who eventually wrote a biography about Einstein. *In 1933, Ilse died of tuberculosis*.

❧ III ☙
EINSTEIN HAD A "MIRACLE YEAR"

"Imagination is more important than knowledge. Knowledge is limited. Imagination encircles the world."
ALBERT EINSTEIN

☙❧

When Einstein graduated in 1900, he faced possibly one of the biggest crises he had ever suffered. All this time he had studied and advanced at his "own" rate. He had always taken the liberty at cutting classes when he wanted to, and this angered some of his professors.

☙❧

One such professor happened to be **Weber** who Einstein needed a reference/letter of recommendation and Webber flat out refused to give him one because he had skipped so many classes. Due to this fact, every position that Einstein applied for, he was turned down. Einstein took the attitude that it was not fair and how could he have had a part in his bad luck. Imagine that; it sounds like Einstein had issues in taking responsibility, does it not?

※

It was years that Einstein harbored the thought that his father had died thinking that he, Albert, was a failure. This thought would cause Albert extreme sadness when he would think about it. How could a man let his father die thinking his son was a failure? If only Einstein could have known what was in front of him?

※

Albert had learned while at school that light speed is the same, so it does not matter how fast anyone moves. It made Newton's laws of motion a bunch of bolognas since Newton's Law had no absolute velocity.

※

All this said and thought about made Einstein start to formulate a new principle of relativity: *Einstein's Principle:"the speed of light is a constant in any inertial frame (constantly moving frame)."*

※

It was during 1905 that people often refer to as Einstein's **"Miracle Year"** because he came out with four published papers and every one of them would alter the course of our world and not just modern physics. They addressed fundamental problems about space, time, motion, matter, and the nature of energy.

※

It was a viewpoint that was concerned with the transformation and production of "**Light and the Quantum Theory**" that was used by Einstein so he could explain the photoelectric effect. He explained that if light occurred in what we now know as **"photon"** or tiny packets, it should be able to take out the electrons found in metal.

※

Einstein that same year offered for the first time experimental ***"proof that atoms"*** did exist. He did so by analyzing tiny particles that showed motion (Brownian movement) suspended in a glass of still water. By this finding, he had captured the ability to calculate sizes of jostling atoms known as "*Avogadro's number*." It came to be known as the "*Movement of Small Particles Suspended in Stationary Liquids Required by Molecular-Kinetic Theory of Heat.*"

※

When it came to the "*Electrodynamics of Moving Bodies*," Einstein was on fire when he laid out his mathematical theory of "***special relativity***." With Special Relativity it can see the light as a field of continuous waves and not particles.

※

Albert's brain was whirling overtime as it came out with "*Does the Inertia of a Body Depend Upon Its Energy Content?*" Which almost got submitted as an afterthought. It was almost not presented, and that would have been a tragedy with him being so "on fire." The theory he almost forgot to submit revealed that the relativity theory was what led to the famous equation "$E=mc^2$."

It gave us our first way to explain the source of energy from the Stars and the Sun. This very equation could predict the evolution of power with almost a million times increased efficiency than that which was obtained by the ordinary physiochemical methods. Even at first, Albert did not get the full picture of the implications of his formula, even though he was able to suggest the heat that was produced by radium could mark conversion of tiny amounts of the bulk of radium salts into energy.

Albert decided that he would submit a paper for his doctorate in 1905. Why not? It was such a Miracle Year, let's keep going!

Sure, there were other scientists out there like Lorentz and Poincare who had bits and pieces of the special *relativity theory*. But, it was Albert who held the Golden Key to the entire theory, and it was he that realized it was a universal law of nature, not some figment of the imagination or motion in the ether as the other two scientists wanted to believe.

❦

Albert wrote in one of his letters to Mileva as he referred to "our theory," has made some wonder that maybe it was she who was the co-founder of the *"theory of relativity."* Since Mileva had failed her graduate exams twice, she had decided to quit physics altogether. So, whether she helped or not will always be a mystery.

❦

It is essential not to forget that during the 19th century there were two islands of thought when it came to physics: There were Newton's laws on motion, and there were Maxwell's ideas about light. Einstein had to take it upon himself in realizing and proving they contradicted each other and one of them had to be wrong.

❦

It seemed at first that everyone was ignoring what Albert had to say in his papers. But, it all started changing when a physicist by the name of Max Planck who had discovered the quantum theory took notice.

❦

It was in 1907 when Albert went face to face with the problem of gravitation. He started working with crucial insight into the fact that: "***gravity and acceleration are equivalent***," they were two angles of the exact phenomenon.

❦

Albert even had minor work that resonated with the world. In 1910, he answered an age-old question as to "Why is the sky blue?" He wrote a paper explaining this phenomenon that was called "critical opalescence." It solved the problem as you examined the summed effect of the light scattering by its molecules in our atmosphere.

※

In 1915, The General Theory of Relativity, which had taken Albert eight years working on the issue of gravity. In Albert's general relativity theory, he shows that ***energy and matter will mold into the shape of the space and the flow of time***. What we can feel as gravity and its "force of pull" is simple. It is a sensation of following the most direct path through tortuous, four-dimensional space-time.It does sound like a radical vision:but space is no more in the same box than the universe comes in; what we find instead is: time and space, energy and matter are, as Einstein can prove, all locked in a most intimate embrace once you understand.

※

It was shortly after Max Planck had commented on Albert's accomplishments that Einstein was being asked to speak at meetings internationally, like the Solvay Conference, and from there he rapidly rose in the world of academia.

※

Talk about a Year of Miracle! He was offered positions at institutions of prestige such as the Swiss Federal Institute of Technology, University of Prague, University of Zurich, and the University of Berlin where he sat as the Director for the Kaiser Wilhelm Institute for Physics for twenty years.

※

Albert's fame was spreading. He lived on the road, giving speeches at international affairs and lost in his deep thoughts of relativity.

❧ IV ❧
EVEN A GENIUS MAKES MISTAKES

"Try not to become a man of success, but rather try to become a man of value."
   ALBERT EINSTEIN

※

From 1905 to 1915, Albert was consumed with this nagging thought that there was a crucial problem, a possible flaw in his theory: he realized it had never made mention of acceleration or gravitation. He had a friend by the name of Paul Ehrenfest that had noticed if the disk is spinning, the outside rim will travel faster than it does in the center, so by special relativity you could place meter sticks on its circumference, and they should shrink. That would explain the Euclidian plane geometry had to fail for the disk.

※

It would obsess Einstein's thoughts for the next ten years as he tried to develop **a theory of gravity** regarding when it came to the curvature of space-time. To Albert, Newton's idea of gravitational force seemed to be a by-product of a something much more profound: the bending of time and space.

Then came November 1915, Albert could finally take a deep breath as he felt he had completed the *"general theory of relativity,"* which would leave a mark on the world as his masterpiece.

During the summer of that year, Albert gave two-hour lectures, six times at the University of Gottingen to thoroughly explain the version that he felt was complete on *general relativity*.

Then along came a real smack in the head when David Hilbert, a mathematician, that had put together the lectures for Albert at the University wrote a paper in November on the subject of *general relativity* five days before Albert. David Hilbert acted like it was indeed his work. He deserved to be tarred and feathered and driven out of town.

Albert and Hilbert did later patch up their quarrel and

remained friends. It seemed from letters that it was Einstein who put his right foot forward first.

> *He penned a letter to Hilbert saying: I struggled against a resulting sense of bitterness, and I did so with complete success. I once more thought of you in unclouded friendship and would ask you to try to do likewise toward me.*

Even today, some physicists refer to this action that the equations are obtained as the Einstein-Hilbert action, but, they know it solely belonged to Einstein.

SOMETIMES, THE MISTAKES COME BACK TO HAUNT THE FAMILY

༄༅

There is a girl, a granddaughter that died by the name of **Evelyn Einstein**. She was an adopted child at birth in 1941 by Hans and Frieda Einstein, Albert's son, and daughter-in-law.

༄༅

Evelyn said as a child that her parents told her that her real birth parents were her grandfather Albert Einstein and a ballet dancer with whom he had an affair.

༄༅

Evelyn only got to see her "grandfather" infrequently after her family moved to California and grandpa Albert lived in Princeton.

༄༅

Evelyn had no proof of anything of the sort, but in all the interviews she was subjected to she told the same story. She had been raised by Albert's son to save the entire Einstein family from being embarrassed.

※

In another interview that she agreed to, she said: "I am outraged. It is hard for me to grasp that I would be treated as I have, which has been terrible."

※

At one time, Evelyn had been married for 13 years to Grover Krantz, an eccentric anthropology professor from America who had become famous while trying to prove that *"Bigfoot"* did exist.

※

After her marriage was over, she seemed to hit bottom, and became a 'dumpster diver,' looking for her next meal.

※

When she died, she left behind no survivors. Money does not buy everything if you are never given any part of it or remembered for who you are in the family.

※

Evelyn had been suffering for several years with diabetes,

heart problems, and cancer. Evelyn said that Albert was never any "great being of science," to Evelyn he was just a plain old grandpa.

❧

Evelyn Einstein was intelligent; *she spoke five different languages*, and in Medieval Literature, she had a Master's degree. All she had done with her life was worked as a police officer, cult deprogrammer, and an animal control officer.

❧

Evelyn Einstein was homeless and had to live in her car for months after she went through a bitter divorce and died when she was 70 in Albany, California.

❧

Albert Einstein, one of the world's genius is considered by some to have been a, well, "pervert." Maybe one should say he enjoyed sexual conquests and thought about it all the time.

❧

I guess what caused him to be looked upon by some as a pervert was because he was so brazen about it with his wives. At one time he would have six girlfriends and would tell his wife how they were always showering him with what he called "unwanted" affection, according to some of his letters.

❧

It is hard to believe that it was "unwanted" affection on his part since he bragged about it so much. He wore all of it like a badge of honor, never giving it a second thought that someone may only be interested in him for his money or his fame and that in truth, his wrinkled clothes and unkempt appearance for most women would be a completeturn off?

Albert spent little time at home, but he was continually writing letters to his family. He would tell them all about his day, his discoveries and the people he had met and the new things he had seen.

The letters that were released in the 80s prove that when he was married to his second wife, who was his first cousin, he was cheating on her with his then secretary, Betty Neumann, and many others we are sure.

In his letters, Albert would tell about six women that he would spend time with and those who would bring him presents while he was married to his second wife.

These letters did not come to light until 1986 as his step-daughter Margot requested that they not be released until she had been dead for 20 years. One could not blame her.

Einstein identified that some of the women included an Ethel, Estella, a Toni and his "Russian Spy Lover," Margarita. He refers to some of the others only by their initials, such as L and M.

❈

Albert said in the letters that it was true M. had even followed him to England and the fact that her stalking me is getting way out of control. He went on to tell Margot that of all his dames, he was only attached to Mrs. L, and she is decent and harmless.

❈

He sent Margot another letter, and Albert asked her to pass on a note to Margarita, to stop providing inquisitive eyes with little tidbits.

❈

Barbara Wolff of the Hebrew University revealed that the M was a Berlin debutante *Ethel Michanowski* and she and Albert were involved during the 1920's and '30s. Wolff said it was an affair, but the only other information she would reveal was that M. was 15 years younger than Albert and was very helpful to his stepdaughters.

❈

In one of his letters to Elsa, he even opened up and said he thought that he was about fed up with his *"theory of relativity"* because when you are involved in something like that so much, even that interest will fade.

❧

The newly found letters also tell the real story about Einstein's money from **the 1921 Nobel Prize**. There was that niggling little term of his divorce that the entire sum would be deposited into a Swiss bank account, and Maric could draw on the interest for her, Eduard, and Hans Albert to live on.

❧

But what did Albert do? He invested most it in the United States and then lost most of it in the Depression. Go, Einstein! And, I mean GO Einstein! A man who has an IQ of 160 and does not evaluate the monetary situation before his investment, does leave one wondering. Did he not have a financial advisor? Why did he not obligate his agreement with his first wife?(This is the second version found in the research.)

❧

It made Maric mad, and she felt betrayed, and she had every right to feel that way as Albert did not stick to his part of the bargain making her continually have to ask him for money.

❧

In the end, he did give her more money than he had won with the Nobel Prize. The prize itself, in 1921, was worth $28,000, and in today's money, it would be $280,000.

❧

Today, Albert's womanizing is over, but his image and name still draw almost ten million dollars a yearand all that goes to research and scholarships at the Israeli University.

❧ V ☙
EINSTEIN FINDS HE HAS ENEMIES

"Unthinking respect for authority is the greatest enemy of truth."
ALBERT EINSTEIN

❦

Einstein was a deep thinker, and that was no doubt. He was convinced that the theory of *general relativity* was right because it was so beautiful, and it could accurately predict Mercury's orbit around the Sun.

❦

In 1916, Einstein wrote and published *"Relativity, the Special and the General Theory: A Popular Exposition,"* 1920.

❦

His theory could predict by measure a deflection of light all around the Sun. Because of that fact, he offered to fund an entire expedition so that they could measure the deflection of starlight in the throes of a Solar eclipse. But Albert's work was to be interrupted by something called World War, and he was angry. Disgusted about the entire war, he called it *"the measles of mankind."*

※

He would go on to write, "At such a time as this, one realizes what a sorry species of animal one belongs to."

Einstein felt that racism of any kind was a disease.

※

After the war, chaos reigned in November 1918, and radical students took control of the University of Berlin, grabbed the college rector, several professors, and took them hostage. The college was worried that if they called the police, there would be a tragic result. But since Einstein was so respected by both faculty and students, he was the logical candidate to be called in as a mediator for the crisis. Einstein and Max Born did broker a compromise and resolved the issues.

※

Since Albert had always been a proponent of civil rights, he had objected to the way African Americans were treated. Einstein himself had suffered anti-Semitic discrimination in Germany before World War II, had worked with several civil rights activists and organizations and demanded they denounce segregation and racism and demanded equality.

❧

There was an African-American civil rights supporter and singer by the name of Marian Anderson who was not being allowed a room at hotels and was being stopped from eating in public restaurants, so Einstein invited her to his house.

❧

A bloody race riot broke out in 1946, and 500 state troopers carrying automatic weapons attacked and virtually destroyed every business owned by a black citizen within four blocks in Tennessee. 25 black men were jailed for attempted murder. Albert joined with Langston Huges, Thurgood Marshall, and Eleanor Roosevelt to fight for justice for the men. 24 of those 25 men were acquitted.

❧

When two black couples in Monroe, Georgia were found murdered, and there was no justice served, Albert was furious; he wrote a letter to President Truman asking for prosecution of the lynchers and passage of a federal law for anti-lynching. Albert became friends with the actor Paul Robeson, and when he became blacklisted due to his activities against racism, Albert again opened his house to a friend of 20 years.

❧

Einstein did not forget what he had wanted to do before the war. So as soon as he could, after the war, two expeditions left to test Albert's theory and his prediction about deflected starlight when near the Sun.

❦

One of his groups sailed to the Island of Principe, which lay off the coast of Africa, while the other group was sent to Sobral in northern Brazil to observe the upcoming eclipse of the Sun on May 29, 1919.

❦

Later that year on November 6th, the results were finally revealed in London to a joint meeting of the Royal Astronomical Society and the Royal Society itself.

❦

The newspaper "The Times" of London headline read, "Revolution in Science – New Theory of the Universe – Newton's Ideas Overthrown – Momentous Pronouncement – Space 'Warped." Immediately, Albert Einstein was a household word and a world-renowned physicist.

❦

The entire world wanted a part of Einstein, and in 1921, he started the first of many world tours in France, Japan, England, and even the United States.

❦

It did not matter where he went, the crowds were wild and numbered into the thousands. While on his way to Japan he heard he had been given the Nobel Prize for Physics, but it was not for his relativity theory. Instead, it was for his photo-electric field. So, when he accepted the Nobel Prize, he talked about relativity. That would show them.

※

Einstein went so far as to become involved in cosmology. With his equations, he felt he could predict that our universe is dynamic – contracting and expanding. It was a direct contraindication to what the previous view had been that the universe was static.

※

In 1928, Einstein was so overworked that he developed an enlarged heart that took him almost an entire year to recover from and be able to get back to his work.

※

Einstein was able to find some recreation in playing his grand piano. A lot of his leisure time was also spent playing his violin. He would play in trios or quartets with his friends who were musically inclined to the fun.

※

In 1929, after Albert turned 50 years old, he decided to build a summer house in Caputh where he would live with his little family every year starting in spring until the late part of autumn. He would do this until 1932.

※

In 1929, the astronomer Edwin Hubble identified that our universe was expanding, and this confirmed Einstein's work. Einstein made a visit in 1930 to Mount Wilson Observatory in California.

☙❧

Hubble and Einstein met, and Albert told Hubble that the *cosmological constant* was his **"biggest mistake."** There is satellite data; recent that is, that has shown that this cosmological constant is probably not zero as thought. It seems to dominate the energy-matter content of our entire universe. It seems that Albert Einstein's "blunder," is the determinant of the ultimate fate of our universe.

☙❧

At this period of his life, Einstein started corresponding with other "thinkers" that were influential during that time. Sigmund Freud (he and Einstein both had boys that had mental difficulties), and an Indian Mystic.

☙❧

Albert started clarifying his views on religion, saying that he had the belief there was an "old one" who was the ultimate lawgiver. He went on to write that he, himself did not believe there was a personal God that would intervene in human affairs but did believe in the 17thcentury Dutch Jewish philosopher God.

☙❧

To this Dutch Jewish philosopher God Albert said, was the

God of beauty and harmony. Albert believed his task was to make up a master theory that allowed him to be able to "read God's mind."

※

Albert wrote,
I am not an atheist, and I do not think I can call myself a pantheist. We are in the position of a little child entering a huge library filled with books in many different languages. The child dimly suspects a mysterious order in the arrangement of the books but doesn't know what it is. That, it seems to me, is the attitude of even the most intelligent toward God.

※

All of Einstein's fame and theories, of course, led to a backlash. The backlash movement (Nazi) targeted "*relativity*," and branded it "Jewish physics," and they sponsored book burnings and conferences so that they could denounce Albert and all of his theories.

※

The Nazis didn't stop there but enlisted other Nobel laureates and physicists such as Philipp Lenard and Johannes Stark to try and make Einstein look like a fool.

※

In 1931, a book "*Authors Against Einstein*" was published.

When Albert was asked to comment on the book and what it had said about him and his theory of relativity by all the different scientists, Einstein simply replied, "to defeat relativity one did not need the word of 100 scientists, just one fact."

※

In December 1932, Einstein without looking back left Germany forever. He was moving and would never go back. It was evident to him that his life was in danger if he stayed or even visited there.

※

A magazine had been published by a Nazi organization with Einstein's picture that had the caption across it that said "*Not Yet Hanged*" on the front cover. Albert found out that there had been a price placed on his head. Due to the size of the threat, Einstein had to re-examine his views on pacifism.

※

Einstein came to America and settled himself in Princeton, New Jersey at the new Institute for Advanced Study. It had become the mecca for other physicists all over the world.

※

Coming to America had been a challenge as well. When our State Department got word that Albert Einstein was coming to America to live, they proceeded to grill Albert on what his political views were at that time. Finally, Einstein had a meltdown.

His usual kind face became stern and his familiar melodious voice loud, he cried out: "What is this, an inquisition? Is this an attempt at deception? I do not plan to answer such silly questions. I did not ask to go to America. Your people invited me; yes, begged me. If I am to enter your country as a suspect, I don't want to go at all. If you don't want to give me a visa, then say so. Then I will know where I stand."

❦

As soon as the press got their hands on this information, the State Department issued visas for Elsa and Albert the next day. They started their journey on December 10, 1932, for the United States and arrived at their destination on January 12th, 1933. It was only a couple of weeks later when Hitler took over Germany. It caused the Einsteins' to stay in the United States permanently.

❦

Newspapers trumpeted that the *"Pope of Physics"* had finally left Germany and Princeton was now his new Vatican.

❦ VI ❦
ALBERT HAD HIS PROBLEMS TOO

"Great spirits have always encountered violent opposition from mediocre minds."
ALBERT EINSTEIN

☙❧

During the 1930's, Einstein faced some challenging years of his life. He suffered a lot of personal sorrow. His son, Eduard, suffered a nervous breakdown and was diagnosed with schizophrenia. As sad as it was, Eduard would remain in an institution for the rest of his life.

☙❧

Einstein had a close friend, another physicist by the name of Paul Ehrenfest, who had worked with him on the *general relativity theory* had become increasingly depressed and killed his fifteen-year-old son with Down's syndrome and then turned the pistol on himself in 1933. The reason he killed himself was not to be known until years later when his letters that he never mailed were discovered.

※

Paul Ehrenfest's depression was so tremendous and the burden of his son's mental challenges and the cost of the institution's fees that would be left for his daughters to have to work all the time to pay were all too much, and Paul just snapped right there in the visitor's room of the institution. He pulled out a gun, killed his son and then turned the gun on himself.

※

Einstein's second wife, Elsa, traveled to Paris to care for her dying daughter Ilse who had cancer. Her other daughter, Margot, came back home to be with her mother because Elsa, herself, had become very ill. She had developed kidney and heart problems, and on December 20, 1936, she died in the Einsteins' new Princeton home they had built in 1933.

※

In the late 1930's, physicists were all looking to consider if, infact, their equations might be able to make an atomic bomb. Even in 1920, Einstein had considered making one but then decided against the idea. However, he left the design on the table if there could be a method found to multiplythe power of the atom.

※

It finally happened in 1938-39 when Lise Meitner, Fritz Strassman, Otto Hahn, and Otto Frisch were able to show the vast amounts of energy one could release by splitting the atom of uranium. The news of this possibility created a fire in the physics society.

※

It was in July 1939 when other physicists convinced Einstein he should be the one to write a letter to the United States President Franklin D. Roosevelt encouraging him to go ahead with developing an atomic bomb.

※

Roosevelt wrote Einstein back, of course, letting Einstein know that he had in fact appointed a Uranium Committee to look at this issue.

※

Einstein kept his Swiss Citizenship but became a man of two countries when he became an American Citizen in 1940.

※

It was during the war that colleagues of Einstein's were asked to meet in Los Alamos; New Mexico to be a part of the Manhattan Project and make the first atomic bomb. It seemed to be one of the greatest ironies of Albert's career for the fact that Einstein the pacifist, through this one action, would help start an era of nuclear weaponry to the use he had always opposed. They never asked Albert to be a part of the effort even though it was his equation that launched the initiative.

❦

There are thousands of FBI files; they are now declassified, which explains the reason that Einstein was not a part of the making of the bomb. Albert being a part of the socialist and peace organizations, being the first and foremost reason for not allowing him to be part of this Project.

❦

J. Edgar Hoover wanted him to be tossed out of the United States entirely and use the Alien Exclusion Act to do so, but the U.S. State Department overruled Hoover and Albert remained in the United States.

❦

Hoover had all kinds of agents spying on poor Einstein. Hoover feared that intellectual, left-wing, pacifist Albert might be some threat to the U.S. establishment or who knows, maybe a Soviet Spy! Hoover even had them going through Albert's mail, listening in on his phone calls, and going through his trash for over 20 years. What a waste of taxpayers' money.

❦

Instead of being asked to leave, during the war, Einstein helped the U.S. Navy in evaluating some of their designs for weapon systems of the future. Einstein even aided the war effort monetarily by allowing some of his manuscripts to be auctioned.

※

One handwritten copy of his paper written in 1905 on *special relativity* sold for $6.5 million. It can now be seen in the Library of Congress. 6.5 million dollars was a lot of money during that time of the world's history.

※

When the atomic bomb was dropped on Japan, Einstein happened to be on vacation. When he heard the news, he joined the effort of bringing the use of the bomb under control by forming the "Emergency Committee of Atomic Scientists."

※

The director of the atomic bomb project, Robert Oppenheimer had his security clearance taken away as he was suspected to have leftist associations.

※

Then, Einstein started backing Oppenheimer and did not like the thought of developing the hydrogen bomb. He called for controls internationally on the technology of nuclear bombs. Einstein became ever more drawn into "anti-war activities and more and more involved in rights for the African American."

※

It was in 1952 that David Ben-Gurion, the Premiere of Israel offered Einstein to be the President of Israel. Albert said that he was deeply moved by the fact that they had offered him this position, and at the same time he was ashamed and saddened that he could not accept. He went on to say that his entire life he had been dealing with scientific type matters, and he felt that he lacked what it took in natural aptitude and what it took to deal with people and to handle official functions. These facts alone are enough that I cannot fulfill what is required of this high office, even if I were getting older and my strength wasn't fading. My relationship with the Jewish nation is my strongest human bond, especially since I have become aware of how precarious it's situation is among all the countries of the entire world.

❧

Einstein had a lot to say about God and his Hebrew heritage, and at times he made you think he did not believe in a God of any form.

> *Here is one of his quotes that is quite thought-provoking, "As a child, I received instruction both in the Bible and in the Talmud. I am a Jew, but I am enthralled by the luminous figure of the Nazarene... No one can read the Gospels without feeling the actual presence of Jesus. His personality pulsates in every word. No myth is filled with such life."*

❧

Einstein loved the water and sailing, but he was not very good at it – he had neighbors on Long Island that could have told you how many times they had to rescue him when his sailboat would capsize. He had named the boat "Tinef" (a Yiddish word meaning "worthless"). As worthless as it was he sure liked playing with the sailboat and even worse, Albert never had learned how to swim! It was a good thing his neighbors watched out for him.

※

And Einstein had a bad habit that did not help his health issues. He loved to smoke. It was in 1950 that he accepted a lifetime membership in the Pipe Smokers Club of Montreal. Einstein felt that smoking a pipe helped one to calm down and have objective judgment in human affairs.

※

The doctors' placed smoking bans on Einstein all the time, but they did little good. When Einstein would walk around the Institute for Advanced Study at Princeton, Albert would pick up cigarette butts off the street and take out the discarded tobacco and fill up his pipe. At first, he walked across the meadow to the institute, but he found the street to have more leftover tobacco. Albert always wanted to get up the courage to defy the doctors' bans, but he didn't want to offend his friends.

❦ VII ❦
DID EINSTEIN HAVE A 3RDSON? YOU DECIDE

Not everything that can be counted counts and not everything that counts can be counted.
ALBERT EINSTEIN

☙❧

There was a weird letter which happened to turn up at the Newspaper of "Evening Prague," one of Prague's largest, most read, daily papers.

☙❧

There was in the paper a note that stated: "I would like to thank you and the country on behalf of the Einstein family," is how the handwritten note started, "for the celebration of the anniversary of my father's 100th birthday. I am very grateful."

༺❀༻

It had been signed by a man who said he was the only living child of Einstein. The letter at first was thought to be a joke or was it just another attempt for someone to try and immigrate to the United States, so it was dismissed.

༺❀༻

1980 rolled around, and all of Einstein's children had passed away. No one connected to the famous Mr. Albert Einstein, renowned physicist, who had lived in the city of Prague since 1912 when Albert himself had left for a more prestigious position in Zurich.

༺❀༻

Well, at the least it seems to be that it is what everyone seemed to think before Ludek Zakel Einstein came to the surface with a handful of papers and documents, a story of babies being switched at birth and a sworn statement of the dear woman who had raised him that he was Albert Einstein's son. He is about 63 years old and what else, but a physicist and he is working on projects basedon research from Albert's work. He looks so much like Einstein and has a temperament so like the man who had the ideas of our modern universe. Ludek has a young son that looks just like Albert.

༺❀༻

Zakel stated in a lengthy interview in his apartment that he may never be able to prove he is Albert's son, but he knows that he is. He goes on to say that he can't profit from the fact that he is Einstein's son and that he could only lose money. It would not change the course of hislife.So why would I want to lie about it?

❦

Ludek Zakel has worked all his life in Prague as a scientist, working to become a physicist way before 1972 when he states, Albert's step-daughter sent someone to inform him that his biological mother, had been Elsa. He had been raised by the woman who he had called mother because her child died the day before Ludek was born.

❦

Ludek's spare time is spent sculpting busts of Einstein and painting with watercolors of the same man as he tries to work out in his head the "significance of gravitational waves in space.

❦

He says he is more than ready to submit to DNA testing to prove what he is saying.

❦

This story that he keeps telling is all based on written but not authorized statements of two nurses that of course have been dead for years. There is the solemn vow of Margot, a possible half-sister, whose mother was Elsa; and the signed statement of 93-year-old Eva Zakel who will not speak about the whole matter anymore.(Elsa and Margot are both now deceased.)

❦

It is said that Mrs. Zakel did give birth to a baby that died on April 14, 1932, and she switched the child with the baby boy of Elsa Einstein who was 50 years old at the time. Elsa had come to Prague to see a doctor because she thought she had some tumor, but not pregnant.

※

It seems that Zakel has an answer to every question that anyone can ask. Elsa had told her friends that she didn't want to find medical help there in Berlin at that time in 1932 because of the Nazis rising to power and Einstein and she was about to flee to America.

※

The truth was, Albert did not want any more children, Elsa told her friends. Mrs. Zakel tells in her account that her baby boy died at birth and she was so desperate to make her husband happy because he wanted children. The original name picked out for the Zakel baby had been Jindrich, and in the hospital's log it is scratched out and replaced with Ludek. There has been no documentation ever found in any of the hospital logs of Elsa Einstein.

※

Things like this could not happen today with tagging systems in hospitals and tagging parents to match their babies. But it was very possible 63 years ago.

※

The documents in existence – a birth certificate and the

baptismal records that were re-issued at the time of the Communist era that says Ludek is the son of Albert Einstein – well that could have been a clever ruse to escape their country.

※

Zakel had applied with the American Embassy more than once for citizenship. Every time he was turned down.

※

It will probably forever remain a mystery to the world since finding out Zakel has so much to lose if he is an Einstein. He has inherited buildings and land and if it should be proved he was Albert's son he could lose them all.

VIII
THE END IS SOON TO COME

"Everybody is a genius. But if you judge a fish by its ability to climb a tree, it will live its whole life believing that it is stupid."
ALBERT EINSTEIN

☙❧

Einstein worked to develop more critical ideas towards his theory of *general relativity*. For instance, he wanted to prove things like wormholes, the possibility that there could be time travel, higher dimensions than we had ever dreamed, the creation of the universe, the existence of black holes and he seemed to become more isolated from the rest of the physics community all the time.

☙❧

The other physicists were working more on quantum theory, not relativity.

> *Einstein would often say of their ideas, "God does not play dice with the Universe."*

※

In 1935, it would happen to be the most celebrated in Albert's career on the quantum theory that led to the Einstein-Podolsky-Rosen "thought experiment."

※

Under quantum theory, other individual circumstances could have two electrons that were separated by enormous distances could still have linked properties, like an umbilical cord.

※

With these circumstances, if one were to measure the properties of the first electron, one would know the state of the second electron faster than light speed. Albert concluded that this violated relativity.

※

There have been experiments since that time that have been able to prove that in fact, the quantum theory was correct and not Einstein.

※

It seems that Einstein became more detached from his colleagues because he became so obsessed with discovering a theory to unify forces of our universe. It caused him in later years to stop opposing quantum theory and try to use it alongside gravity and light.

❦

Einstein became so set in his ways. He quit traveling and just took long walks around the grounds of Princeton with some of his closest friends and associates with whom he could discuss religion, politics, unified field theory, and physics.

❦

He was a regular sight to be seen walking around campus. His hair was always in disarray as if he never combed it or had just gotten out of bed. His pants were still wrinkled, and he usually wore a sweater. He never wore socks; they were a bother to him in the area where the big toe is located as they always had a big hole in that spot. Albert's second toe was much longer than his big toe next to it.

❦

It was in 1950, he wrote and published an article in the Scientific American on his theory on strong force, but it was neglected by most, and it remained incomplete.

❦

Albert seemed to believe that you must stand for what you believe in or you will fall for anything. He would never waver if his conscience told him to take action on a matter even if it was unpopular.

⁂

One of these occasions was January 12, 1953, when Albert penned a letter to then-President Harry Truman.

> *It said: "My conscience compels me to urge you to commute the death sentence of Julius and Ethel Rosenberg."*

The two convicted atomic spies were executed five months later.

⁂

Einstein's health had not always been so good. When he was 69, he went to his primary care physician and told him he had been having a lot of pain in the uppermost part of his stomach. He told him that off and on for several years he had been having attacks in his upper abdomen that would last at times for two to three days at a time and he almost always vomited when he had these attacks.

⁂

He went on to add that this seemed to happen about every three to four months. He did smoke a pipe and seemed to be a bit overweight but not that much. When the doctor examined him, he could feel a mass deep down in the center of his stomach that was pulsating.

⁂

Dr. Nissen, the very doctor who had developed the operation known as the *"Nissen Procedure,"* that prevents gastro-esophageal reflux, operated on Einstein by exploratory laparotomy at the Jewish Hospital in Brooklyn. When he opened Albert up, there lay an aortic aneurysm the size of a grapefruit.

☙❧

Nissen knew he could not ligate this large aneurysm and replacing the aorta with a graft was out of the question and yet to be perfected. All he could try to do was to reinforce the wall of the aorta and try to delay the inevitable rupture that would come.

☙❧

A good tissue irritant that produces marked fibrosis is polyethylene. So, Dr. Nissen while having Einstein open wrapped the anterior visible part of the aneurysm with this cellophane, hoping it would cause an intense fibrous reaction in the tissue, which would strengthen the wall of the aneurysm.

☙❧

Albert recovered in three weeks from the surgery during his hospital stay and returned to Princeton, New Jersey to his home.

☙❧

Off and on Albert would have some occasional back pain and would experience some pain much like gallbladder pain.

❧

April 12, 1955, Einstein started having some pretty severe abdominal pain that got even worse the next day. Albert had a pretty good idea of what had happened and at first, refused to go to the hospital. He finally did go to the hospital because he felt he was a burden there at home. The Chief of Surgery at New York Hospital wanted to resect the aneurysm by a new procedure.

❧

Albert refused the surgery and said that he wanted to go when he wanted. He felt you should not prolong your life artificially. He felt that he had done his share on this earth and it was his time to go. By doing so, he would do it elegantly.

❧

The night before Albert expired he had a view of his little round garden from the bed.

> *The nurse taking care of him asked, "Professor do you think God made the garden?"*
> *Einstein said, "Yes, God is both the gardener and the garden" to which the nurse replied, "Oh, I'd not thought of it that way" to which Einstein replied, "Yes, and I have spent my whole life just trying to catch a glimpse of Him at his work."*

❧

He did, however, leave a piece of writing that happened to end unfinished. It happened to be his last words.

> *In essence, the conflict that exists today is no more than an old-style struggle for power, once again presented to mankind in semi-religious trappings. The difference is that this time, the development of atomic power has imbued the struggle with a ghostly character; for both parties know and admit that should the quarrel deteriorate into actual war, mankindis doomed. Despite this knowledge, statesmen in responsible positions on both sides continue to employ the well-known technique of seeking to intimidate and demoralize the opponent by marshaling superior military strength. They do so even though such a policy entails the risk of war and doom. Not one statesman in a position of responsibility has dared to pursue the only course that holds out any promise of peace, the course of supranational security, since for a statesman to follow such a course would be tantamount to political suicide. Political passions, once they have been fanned into flame, exact their victims... Citater fra...*

࿔

Five days after being admitted to the hospital, Albert Einstein developed labored breathing and breathed his last *at 1:15 AM, April 18, 1955*. What a sad time for the history of the world.

࿔

When he died, his body was moved from the hospital out to the funeral home before being cremated in Trenton. Most were not aware, but his brain did not get cremated, it went missing for many years. The pathologist of the hospital had confiscated it. It was finally found 23 years later by the journalist, Steven Levy who happened upon it pickled in a jar by one **Dr. Thomas Harvey** at Princeton where Einstein had expired.

The rest of his body that was cremated; the ashes were spread around the Institute for Advanced Study in Princeton.

Don't think that the pickling of his brain went to waste. There are recent studies that have revealed there were certain parts of Einstein's brain that were unusually convoluted. Not just that, his parietal lobes were "extraordinarily asymmetrical," and the motor cortices and somatosensory areas were "greatly" expanded in the left hemisphere.

Further studies also show that Albert's brain cells had more of one type of brain cells called "glial" cells than our typical brains.

Einstein's IQ was 160, but one has to wonder with the methods we have now of grading IQ, what his might genuinely rise to be.

IX
CHAPTER 8: WHAT EXACTLY WAS THE LEGACY OF EINSTEIN?

"Look deep into nature, and then you will understand everything better."
ALBERT EINSTEIN

※

It seems as we look back that Einstein was far ahead of his time. His significant piece of unified field theory remained a total mystery in his lifetime.

※

It was not until the 1970s and into the 80s that physicists started to unravel secrets associated with the *strong force* of the *quark model*.

※

Even today Einstein's work allows other physicists to win Nobel Prizes as they build on his theories. In 1993 there was a Nobel Prize awarded to those who discovered gravitational waves that had been a prediction of Einstein's.

❧

Black holes now number in the thousands in space. The satellites we have in space have been able to verify precisely what Einstein spoke of in his lifetime.

❧

Even after Einstein retired, he kept working towards unifying the fundamental perceptions of physics, geometrization, to take the opposite approach, to the bulk of physicists.

❧

Everything that Einstein researched is well cataloged, and his most important works include The Special Theory of Relativity, General Theory of Relativity, Relativity Translations, The Evolution of Physics, About Zionism, My Philosophy, Why War?, Out of My Later Years, Investigations on Theory of Brownian Movement are some of the most important.

❧

Albert Einstein was the recipient of honorary doctorate degrees in medicine, philosophy, and science from universities in America and Europe. He was awarded the Franklin Medal from the Franklin Institute and the Copley Medal of the Royal Society of London.

❧

Einstein was always forward thinking, and he was not afraid to think of what would happen after his death.

❧

Notable for Einstein was that as he was nearing his last years of life, his views of God seemed to change.

> *One notices this in this quote: "As a child, I received instruction both in the Bible and in the Talmud. I am a Jew, but I am enthralled by the luminous figure of the Nazarene. No one can read the Gospels without feeling the actual presence of Jesus. His personality pulsates in every word. No myth is filled with such life."*

One must admit this is in direct contrast as to how he felt when he was a hardened scientist.

❧

In his last will, he stated that everything he had, and all his intellectual property should be placed in a trust guarded by his stepdaughter and his secretary. His violin he left to his grandson. After his trustees had passed away, what remained of the rights of the trust were to be given to the Hebrew University in Jerusalem.

X
CONCLUSION

STRENGTHS

❦

lbert Einstein had an IQ of 160 and who in the world would not love to have an IQ that high?

❦

Albert Einstein was liked well enough by professors and students alike that he could calm a terrible riot on the university where professors had been taken as hostages.

❦

Albert Einstein believed that if you could dream it and you worked toward it that you would eventually succeed in attaining the goal that you were reaching for. He set this example on many occasions.

WEAKNESSES

❧

Albert Einstein gave very little of himself to his family. It is the feelings of this author that he never really knew any of his children nor they him.

❧

Albert Einstein was never good to either of his wives. He was a braggart and boastful when it came to his conquests, and he did not care who in his family knew about the other women.

❧

Albert Einstein preferred being isolated from others rather than having to deal with other customs, opinions, and the prejudices of others.

❧

Albert Einstein, suffering from Asperger's Syndrome as a child, had so much to overcome as a child, and yet he forged on, and that alone was much to be admired. What he was able to do for the entire world was nothing short of amazing.

❧❧❧

As I said in the book when he believed passionately in something and he did on many issues, he stood for what he believed in and stood firm. He did not let anyone talk him out of it.

❧❧❧

He teaches us the reader to never give up on our dreams and that everyone is capable of reaching their goals no matter their humble beginnings.

❧❧❧

Albert has many, many quotes that are highly thought-provoking and great to have nearby for reading.

INTERESTING BOOKS ON ALBERT EINSTEIN YOU MAY ENJOY:

Interesting Books on Albert Einstein you may enjoy:

❧

Einstein's Tears

❧

The Einstein File (From the FBI archives)

❧

The World as I see It (by Einstein)

❧

Albert Einstein – The Human Side

Made in the USA
Columbia, SC
13 November 2018